Major Biological Events

Millions of Years Ago

	Millions of Years Ago	
	0.135	Modern *Homo sapiens* arises.
	2	Genus *Homo* arises.
	4	Australopithecines present.
Gymnosperms, angiosperms widespread. Temperate grasslands, forests expand.	65	Mammals diversify. Primates arise.
Angiosperms arise and diversify.		Major extinction event. Most large reptiles, ancient birds extinct.
		Teleost fish diversify. Dinosaurs dominant. Modern crustaceans common.
Gymnosperms, ferns dominant.		Dinosaur ancestors common. First mammals, birds.
Conifers appeared.		Mammallike reptiles common. Major extinction of invertebrates, amphibia.
Forests widespread. Coal deposits form.		First reptiles. Amphibia diversify. Major extinction event.
First forests. Vascular plants and seeds present.		First insects, sharks, amphibians. Fish diversify.
Green, red, brown algae common.		First land arthropods. Jawed fish arise.
First vascular land plants probably appeared.		Second major extinction event. Jawless fish diversify; large invertebrates present; mollusks diversify. First tracks left by land animals.
Algae dominant.		First major extinction event. Trilobites common; onychophorans, first jawless fish at end of period. Evolution of many phyla.
Algae abundant. Multicellular organisms: algae, fungi. Cyanobacteria diversified. Eukaryotes present: green algae, protists.		Wormlike animals, cnidarians present.
Photosynthetic cells liberate oxygen.		
Origin of life. Prokaryotic heterotrophs. Chemical evolution.		

THE NATURE OF LIFE

ABOUT THE AUTHORS

JOHN H. POSTLETHWAIT, Professor of Biology at the University of Oregon, holds a B.A. degree from Purdue University and a Ph.D. from Case Western Reserve University, and he trained as a post-doctoral fellow at Harvard University. He was Visiting Research Scientist for a year each at the Institut für Molekular Biologie of the Austrian Academy of Sciences in Salzburg, Austria; the Laboratoire de Génétique Moléculaire, Faculté de Médecine, in Strasbourg, France; and the Imperial Cancer Research Fund in Oxford, England. From 1971 to the present, his research in developmental and molecular genetics has been funded by grants from the National Institutes of Health, the American Heart Association, and the U.S. Department of Agriculture, and he has published the results of his work in more than one hundred research articles. He has taught general biology to majors and nonmajors since 1971 at the University of Oregon, where he received the Ersted Award for Distinguished Teaching. His current efforts include work funded by the National Science Foundation program in biology education, developing a workshop-oriented approach in the teaching of non-biology majors that emphasizes open-ended investigation and the application of biological concepts to solving problems in society.

JANET L. HOPSON is a lecturer for the Science Communication Program, University of California at Santa Cruz, as well as a freelance science writer. She has also taught writing courses at the University of California at Berkeley and Mills College. She holds B.A. and M.S. degrees from Southern Illinois University and the University of Missouri. Coauthor of three other biology textbooks for McGraw-Hill, *Biology* and *Essentials of Biology* with Norman K. Wessells, and *Biology! Bringing Science to Life* with John H. Postlethwait and Ruth C. Veres, she has also written a trade book on the human sense of smell. She won the Russell L. Cecil Award for magazine writing from the Arthritis Foundation and has published dozens of articles in national magazines and newspapers, including *Smithsonian, Psychology Today, Science News, Science Digest, Outside,* and others. Her biography is included in the first edition of *Who's Who in Science and Engineering.*

SECOND EDITION

THE NATURE OF LIFE

JOHN H. POSTLETHWAIT UNIVERSITY OF OREGON

JANET L. HOPSON UNIVERSITY OF CALIFORNIA, SANTA CRUZ

McGRAW-HILL, INC.
NEW YORK ST. LOUIS SAN FRANCISCO AUCKLAND BOGOTÁ CARACAS HAMBURG LISBON
LONDON MADRID MEXICO MILAN MONTREAL NEW DELHI PARIS SAN JUAN
SÃO PAULO SINGAPORE SYDNEY TOKYO TORONTO

With appreciation for the love and support of my mother, Sara M. Postlethwait, and my father, Samuel N. Postlethwait, a wonderful inspiration for students and teachers of biology. —J.H.P.

For Samuel F. Smith and Charles R. Billings, with my deepest appreciation and affection. —J.L.H.

The Nature of Life

1 2 3 4 5 6 7 8 9 0 VNH VNH 9 0 9 8 7 6 5 4 3 2

ISBN 0-07-050633-7

Library of Congress Cataloging-in-Publication Data

Postlethwait, John H.
  The nature of life/John H. Postlethwait, Janet L. Hopson.—2nd ed.
    p.   cm.
  Includes bibliographical references and index.
  ISBN 0-07-050633-7
  1. Biology.   2. Life (Biology)   I. Hopson, Janet L.   II. Title.
  [DNLM: 1. Biology.   QH 308.2 P858n]
QH308.2.P67   1992
574—dc20
DNLM/DLC   91-25800
for Library of Congress   CIP

Figure 2.21c: Adapted with permission of Procter & Gamble.

Figure 12.15: Adapted from "The Human Gene Map," compiled by Victor A. McKusick, 1986. Reprinted with permission of Dr. Victor A. McKusick, University Professor of Medical Genetics, Johns Hopkins University, and Cold Spring Harbor Laboratory.

Figure 33.28: Modified from *Horses: The Story of the Horse Family in the Modern World and Through Sixty Million Years of History* by G. G. Simpson. Copyright © 1951 by Oxford University Press, Inc.; renewed 1979 by G. G. Simpson. Reprinted by permission.

Sponsoring Editors: Denise Schanck, June Smith

Senior Associate Editor: Mary Eshelman

Senior Editing Supervisor: Alice Mace Nakanishi

Assistant Production Manager: Pattie Myers

Designer: BB&K Design

Art Coordinator: Marian Hartsough

Developmental Art Consultants: Iris Martinez Kane, Cherie Wetzel, Arthur Ciccone

Illustrators: Martha Blake, Wayne Clark, Cecile Duray-Bito, JAK Studio, Paula McKenzie, Linda McVay, Elizabeth Morales-Denney, Victor Royer, Carla Simmons, Tek-Nēk´ Inc., John Waller, Judith Waller, Cyndie C.H.-Wooley

Photo Researchers: Darcy Lanham, Monica Suder

Copyeditor: Janet Greenblatt

Proofreader: Sarah Miller

Indexer: Barbara Littlewood

Compositor: Graphic Typesetting Service

Color Separator: Black Dot Graphics

Printer and Binder: Von Hoffmann Press

Cover Photos: Rusty parrotfish (*Scarus ferrugineus*) (inset) and rusty parrotfish scales (background) by Jeff Rotman

Cover Photo Researcher: Natalie Goldstein

Cover Color Separator and Printer: New England Book Components

Production assistance by: Betsy Dilernia, Brian Jones, Jane Moorman, Tralelia Twitty

Credits are continued on pages C-1 to C-6.

CONTENTS IN BRIEF

1 The Nature of Life: An Introduction 2

PART ONE
Life's Fundamentals

2 Atoms, Molecules, and Life 24
3 Cells: The Basic Units of Life 52
4 The Dynamic Cell 80
5 How Living Things Harvest Energy
from Nutrient Molecules 100
6 Photosynthesis:
Trapping Sunlight to Build Nutrients 122

PART TWO
Perpetuation of Life

7 Cell Cycles and Life Cycles 138
8 Mendelian Genetics 164
9 DNA: The Thread of Life 188
10 How Genes Work: From DNA to
RNA to Protein 208
11 Genetic Recombination and Recombinant
DNA Research 228
12 Human Genetics 244
13 Reproduction and Development:
The Start of a New Generation 266
14 The Human Life Cycle 290

PART THREE
Life's Variety

15 Life's Origins and Diversity on Our Planet 316
16 Life As a Single Cell 336
17 Plants and Fungi:
Decomposers and Producers 356
18 Invertebrate Animals:
The Quiet Majority 378
19 The Chordates:
Vertebrates and Their Relatives 402

PART FOUR
How Animals Survive

20 An Introduction to How Animals Function 430
21 Circulation:
Transporting Gases and Materials 446
22 The Immune System and the Body's Defenses ... 464
23 Respiration: Gas Exchange in Animals 482
24 Animal Nutrition and Digestion:
Energy and Materials for Every Cell 494
25 Excretion and the Balancing of
Water and Salt 516
26 Hormones and Other Molecular Messengers 534
27 How Nerve Cells Control Behavior 552
28 The Senses and the Brain 568
29 The Dynamic Animal:
The Body in Motion 588

PART FIVE
How Plants Survive

30 Plant Architecture and Function 606
31 Regulators of Plant Growth and Development ... 630
32 The Dynamic Plant:
Transporting Water and Nutrients 646

PART SIX
Interactions: Organisms and Environment

33 The Genetic Basis for Evolution 664
34 Population Ecology: Patterns in
Space and Time 690
35 The Ecology of Living Communities:
Populations Interacting 710
36 Ecosystems: Webs of Life and
the Physical World 730
37 The Biosphere: Earth's Fragile Film of Life 750

PART SEVEN
Behavior and the Future

38 Animal Behavior: Adaptations for Survival 774

CONTENTS

Preface xxv
A Guided Tour to *The Nature of Life* xxx

CHAPTER 1 ■ The Nature of Life: An Introduction 2

To Be Alive: Unifying Themes 4

Life Traits Involving Energy and Organization 4

Life Characteristic 1: Order Within Biological Systems 5
Life Characteristic 2: Metabolism 7
Life Characteristic 3: Motility 7
Life Characteristic 4: Responsiveness 7

Life Characteristics Related to Perpetuation 8

Life Characteristic 5: Reproduction 8
Life Characteristic 6: Development 8
Life Characteristic 7: Heritable Units of Information,
the Genes 9

**Life Characteristics Related to Evolution
and Environment** 10

Life Characteristic 8: Living Things Evolve 10
Life Characteristic 9: Living Things Are Adapted 10

Evolution: Biology's Central Theme 11

Organizing Life's Dazzling Diversity 12
Natural Selection: A Mechanism of Evolution 12

**The Process of Science: A Major System for
Learning About Life** 16

Fundamental Principles: Causality and Uniformity 16
The Power of Scientific Reasoning 17
Testing Generalizations 17

**How Biological Science Can Help Solve
World Problems** 19

FEATURES
■ The Courtship of a Scarlet Frog .. 2
■ Biology: A Human Endeavor
  BOX 1.1: Darwin, Wallace, and
  Evolution by Natural
  Selection 15
■ How Do We Know?
  BOX 1.2: Evolutionary Oddities
  Prove the Rule 18
■ Connections 20
  Highlights in Review
  Key Terms
  Study Questions
  For Further Reading

PART ONE ■ Life's Fundamentals

CHAPTER 2 ■ Atoms, Molecules, and Life 24

Atoms and Molecules 26

The Nature of Matter: Atoms and Elements 26
Organization Within the Atom 27
Variations in Atomic Structure: Atomic Bombs and
Nerve Impulses 28
Electrons in Orbit: Atomic Properties Emerge 29
Molecules: Atoms Linked with "Energy Glue" 30

Life and the Chemistry of Water 33

Water, Temperature, and Life 33
Mechanical Properties of Water 34
Chemical Properties of Water 35
Water Dissociation: Acids and Bases 35

The Stuff of Life: Compounds Containing Carbon 38

Carbon: Compounds and Characteristics 38

FEATURES
■ Water, Survival, and the
  Chemistry of Living Things 24
■ Personal Impact
  BOX 2.1: Magnetic Resonance
  Imaging and the Secrets in
  Hydrogen 28
■ Personal Impact
  BOX 2.2: Free Radicals and
  Cigarette Smoking 31
■ Connections 49
  Highlights in Review
  Key Terms
  Study Questions
  For Further Reading

Carbohydrates 38
Lipids 43
Proteins: Key to Life's Diversity 44
Nucleic Acids: Information Storage and Energy Transfer 48

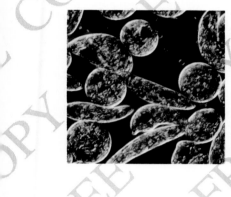

CHAPTER 3 ■ Cells: The Basic Units of Life 52

The Discovery and Basic Theory of Cells 54

The Units of Life: An Overview 55

The Two Major Kinds of Cells 55
The Numbers and Sizes of Cells 59

The Common Functions and Structures of All Cells 61

The Cell's Dynamic Boundary: The Plasma Membrane 61
The Nucleus: The Cell's Control Center 65
Cytoplasm and the Cytoskeleton: The Dynamic Background ... 67
A System of Internal Membranes Providing Synthesis,
Storage, and Export 67
Mitochondria: The Cell's Powerhouse 71

Specialized Functions and Structures in Cells 73

Plastids: Organelles of Photosynthesis and Storage 73
Vacuoles: Not-So-Empty Vesicles 73
Cellular Whips: Cilia and Flagella 74
Cell Coverings 75
Links Between Cells 76

FEATURES
■ *Euglena:* An Exemplary Living
Cell 52
■ How Do We Know?
BOX 3.1: Microscopes: Tools for
Studying Cells 62
■ Connections 77
Highlights in Review
Key Terms
Study Questions
For Further Reading

CHAPTER 4 ■ The Dynamic Cell 80

Cells and the Basic Energy Laws of the Universe 82

The Laws of Thermodynamics 82
Cells and Entropy 83

**Chemical Reactions and Energy Flow in
Living Things** 84

Chemical Reactions: Molecular Transformations 85
Metabolism: Chains of Reactions 85
ATP: The Cell's Main Energy Carrier 87
Chemical Reactions in the Cell: Enzymes Make Them Go 88
Enzyme Form Facilitates Biological Reactions 90

Metabolism: The Dynamic Cell's Chemical Tasks 91

Transport Tasks in the Dynamic Cell 92

Passive Transport, Diffusion, and the Second Law
of Thermodynamics 92
Passive Transport and the Movement of Water 93
Active Transport: Energy-Assisted Passage 95

The Dynamic Cell's Mechanical Tasks 96

FEATURES
■ Red Blood Cells: Activity and
Order 80
■ Connections 97
Highlights in Review
Key Terms
Study Questions
For Further Reading

CHAPTER 5 ■ How Living Things Harvest Energy from Nutrient Molecules 100

ATP and the Transfer of Energy from Nutrient Molecules 102

ATP Structure: A Powerful Tail 102
"Electrical Currents" in Energy Metabolism 103

An Overview of Glycolysis, Fermentation, and Aerobic Respiration 104

Glycolysis: The Universal Prelude 105

Fermentation: "Life Without Air" 106

Alcoholic Fermentation 108
Lactic Acid Fermentation 109
Anaerobic Metabolism: A Final Tally 109

Aerobic Respiration: The Big Energy Harvest 110

The Krebs Cycle: Metabolic Clearinghouse 110
The Electron Transport Chain: An Energy Bucket Brigade..... 112

The Control of Metabolism 117

The Energy Source for Exercise 118

FEATURES

■ A Mountain Hike: Splitting Sugar for Muscle Power 100
■ How Do We Know?
  BOX 5.1: A Yeasty Subject for Research 116
■ Personal Impact
  BOX 5.2: Sarah's Mitochondria.. 119
■ Connections 120
  Highlights in Review
  Key Terms
  Study Questions
  For Further Reading

CHAPTER 6 ■ Photosynthesis: Trapping Sunlight to Build Nutrients 122

An Overview of Photosynthesis 124

The Chloroplast: Solar Cell and Sugar Factory 124

Colored Pigments in Living Cells Trap Light 125

Light, Chlorophyll, and Other Pigments 126

The Light-Dependent Reactions of Photosynthesis 128

Light Excites Electrons 128
Energy Is Stored in Energy Carriers 129
Water Is Split and Oxygen Produced 131

The Light-Independent Reactions of Photosynthesis 131

Fixing Carbon 131
Manufacturing Stable Storage Molecules 132
Regenerating the Starting Compound 132
What Makes Glucose a Better Energy Store Than ATP? 132

A Special Type of Carbon Fixation 133

Photosynthesis and the Global Environment 134

FEATURES

■ The Photosynthetic Champion 122
■ Focus on the Environment
  BOX 6.1: When You See Green, Think of a Green Sea 126
■ Focus on the Environment
  BOX 6.2: Red Algae That Grow in the Dark 129
■ Connections 135
  Highlights in Review
  Key Terms
  Study Questions
  For Further Reading

PART TWO   Perpetuation of Life

CHAPTER 7 ▪ Cell Cycles and Life Cycles 138

Chromosomes: Repositories of Information That Direct Cell Growth and Reproduction 140

Information for Directing Cell Growth: Stored in the Nucleus 140
Genetic Information: Stored in the Chromosomes 141

The Cell Cycle 142

The Cell Cycle in Prokaryotes: One Cell Becomes Two Through Binary Fission 142
Cell Cycle in Eukaryotes: Four Phases of Growth and Division 143

Mitosis and Cytokinesis: One Cell Becomes Two 145

Mitosis: Chromosome Choreography 145
Cytokinesis: The Cytoplasm Divides 149

Regulating the Cell Cycle 151

External Factors Regulating the Cell Cycle 151
Internal Factors Regulating the Cell Cycle 152
Cancer: Cell Cycle Regulation Gone Awry 152

Life Cycles: One Generation to the Next in Multicellular Organisms 152

Asexual Reproduction: One Individual Produces Identical Offspring 152
Sexual Reproduction: Gametes Fuse and Give Rise to New Individuals 153

Meiosis: A Reshuffling and Reduction of Chromosomes 156

The Stages of Meiosis 157
Genetic Variation Arises During Meiosis 157

Meiosis, Mitosis, Sexual Reproduction, and Evolution 161

FEATURES

▪ A Week in the Life of a Small But Famous Wound 138
▪ Focus on the Environment
   BOX 7.1: Radiation Sickness: A Disease of Mitosis 150
▪ How Do We Know?
   BOX 7.2: The Whole Worm Catalog 155
▪ Personal Impact
   BOX 7.3: Down Syndrome: How Mistakes in Meiosis Alter Development 157
▪ Connections 162
   Highlights in Review
   Key Terms
   Study Questions
   For Further Reading

CHAPTER 8 ▪ Mendelian Genetics 164

Genetics in the Abbey: How Genes Were Discovered and Analyzed 166

The Critical Test: A Repeatable Experiment with Peas 167
Mendel Disproves the Blending Theory 167
Segregation Principle for Alleles of One Gene 168

Inheritance of Two Independent Traits 175

Ratios Reveal Independent Assortment 175
Mendel's Results Ignored 176

Geneticists Locate Genes on Chromosomes 176

Different Sexes—Different Chromosomes 176
Sex and White Eyes 177
Mutations Reveal That Each Chromosome Carries Many Genes 177

FEATURES

▪ White Tigers and Family Pedigrees 164
▪ Biology: A Human Endeavor
   BOX 8.1: Mendel and His Mentors 168
▪ Personal Impact
   BOX 8.2: Hinnies, Mules, and Mendel's Rules 181
▪ Connections 185
   Highlights in Review
   Key Terms
   Study Questions
   For Further Reading

Linkage: Genes on the Same Chromosome Tend to Be
Inherited Together 178
Crossing Over: Evidence That a Chromosome Is a Linear
Array of Genes 179

**Gene Interactions: Exceptions Obscure
Mendelian Principles** 180

Interactions Between Alleles 180
Interactions Between Genes 183
Multiple Effects of Individual Genes 184
Environmental Effects on Gene Expression 185

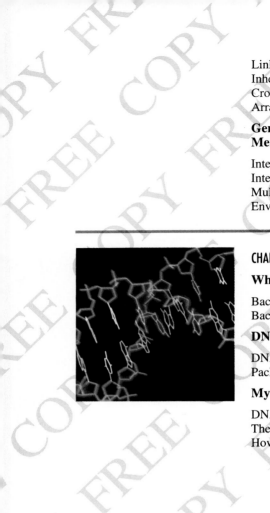

CHAPTER 9 ■ DNA: The Thread of Life 188

What Are Genes? 190

Bacterial Transformation: Evidence for DNA 190
Bacterial Invaders: Conclusive Proof from Viral DNA 190

DNA: The Twisted Ladder of Inheritance 192

DNA: A Linear Molecule 192
Packaging DNA 196

Mysteries of Heredity Unveiled in DNA Structure 196

DNA: A Molecular Template for Its Own Replication 197
The Three Stages of Replication 197
How Does DNA Store Information? 201

FEATURES

■ A Sudden Epidemic and
  Unbeatable Bacteria 188
■ Personal Impact
  BOX 9.1: The Human Genome
  Project 195
■ Personal Impact
  BOX 9.2: DNA Synthesis, AZT,
  and Treating AIDS 200
■ Personal Impact
  BOX 9.3: Antibiotics, Cattle
  Feed, Plasmids, and Your
  Health 205
■ Connections 205
  Highlights in Review
  Key Terms
  Study Questions
  For Further Reading

CHAPTER 10 ■ How Genes Work: From DNA to RNA to
Protein 208

**The Path of Information Flow from DNA
to RNA to Protein** 210

Transcription: Information Flows from DNA to RNA 210
Translation: Information Flows from RNA to Protein 212
Protein Synthesis: Translating Genetic Messages into the
Stuff of Life 215

Gene Mutation: A Change in Base Sequence 216

How a Gene Mutation Changes the Phenotype 218
The Origin of Mutations 218

Regulation of Gene Activity 219

Levels of Control 219
Gene Regulation in Prokaryotes 221
Gene Regulation in Eukaryotes 222

FEATURES

■ Cystic Fibrosis: A Case Study
  in Gene Action 208
■ Personal Impact
  BOX 10.1: Caesar Experiments
  with RNA Synthesis 224
■ Connections 226
  Highlights in Review
  Key Terms
  Study Questions
  For Further Reading

CHAPTER 11 ■ Genetic Recombination and
Recombinant DNA Research 228

Recombination in Nature 230

Recombination: Universal Source of Variation 230
Contribution to Evolution 230

The Power of Recombinant DNA Research 232

How to Construct a Recombinant DNA Molecule 233
How to Clone a Human Gene 233

**Promises and Problems in Recombinant DNA
Research** 234

Reshaping Life: The Promise of Genetic Engineering 235
Recombinant DNA and Environmental Risks 239
Recombinant DNA: Novel Problems of Safety and Ethics 240

FEATURES
■ Building a Bigger Mouse 228
■ Biology: A Human Endeavor
BOX 11.1: Barbara McClintock
and Jumping Genes 231
■ How Do We Know?
BOX 11.2: Manipulating DNA
Molecules by Electrophoresis.. 236
■ Biology: A Human Endeavor
BOX 11.3: Kary Mullis and
Copies of Copies of Copies 240
■ Connections 242
Highlights in Review
Key Terms
Study Questions
For Further Reading

CHAPTER 12 ■ Human Genetics................................... 244

**How Traditional Techniques Reveal Thousands of
Human Genetic Conditions** 246

Human Pedigrees: Analysis of Family Genetic Histories 246
Chromosome Variations Reveal Other Genetic Conditions..... 249

**The New Genetics: A Revolution in the Mapping of
Human Genes** 254

Bypassing Sex: Somatic Cell Genetics 254
Mapping Variations in DNA Structure 255

Helping People with Genetic Diseases 257

Identifying Victims of Genetic Disease 257
Physiological Therapy 259
Protein Therapy 259
Gene Therapy... 259

Preventing New Cases of Genetic Disease 261

Prenatal Diagnosis 261
Detecting Carriers 261
Genetic Counseling...................................... 262

FEATURES
■ Diet Soft Drinks and Human
Genetics 244
■ Personal Impact
BOX 12.1: The Royal
Hemophilia 251
■ Personal Impact
BOX 12.2: The Risks of Genetic
Screening 258
■ Personal Impact
BOX 12.3: DNA Fingerprinting
Can Solve Crimes 260
■ Connections 262
Highlights in Review
Key Terms
Study Questions
For Further Reading

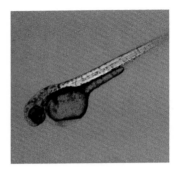

CHAPTER 13 ■ Reproduction and Development:
The Start of a New Generation 266

**Mating and Fertilization: Getting Egg and
Sperm Together** 268

Mating Strategies: Getting Organisms Together 268
Fertilization: The Actual Union of Egg and Sperm 269

Patterns of Early Embryonic Development 271

Cleavage: Formation of an Embryo with Many Cells 272
Gastrulation: Establishing the Three-Layered Body Plan 274
Neurulation: Forming the Nervous System 276

Development of Body Organs: Organogenesis 278

Development of Organ Shape 280
Development of Organ Function: Differentiation 280

Development Continues Throughout Life 281

Continual Growth and Change 281
Cancer: Development Running Amok 284

**The Formation of Gametes: The Developmental
Cycle Begins Anew** 286

Special Egg Cytoplasm Leads to Germ Cells in
the New Individual 286

FEATURES
■ The Omnipotent Egg 266
■ How Do We Know?
BOX 13.1: A Leg Grows from
the Head: The Homeobox 279
■ How Do We Know?
BOX 13.2: How Does a Cell
"Know" Where It Is? 282
■ Connections 287
Highlights in Review
Key Terms
Study Questions
For Further Reading

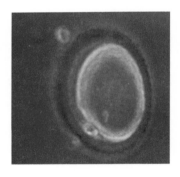

CHAPTER 14 ■ The Human Life Cycle 290

Male and Female Reproductive Systems 292

The Male Reproductive System 292
The Female Reproductive System 295

**Sperm Meets Egg—Or Doesn't: Fertilization, Birth
Control, and Infertility** 298

Fertility Management and the Planet's
Life-Support Systems 299
Birth Control 299
Overcoming Infertility 302

Pregnancy, Human Development, and Birth 302

Implantation and the Chorion: The Embryo Signals
Its Presence 302
Developmental Stages in the Human Embryo 304
Sex Differentiation: Variations on One
Developmental Theme 304
Fetal Life: A Time of Growth 306
Mother's Contribution to the Fetal Environment 306
Crossing the Threshold: The Magic of Birth 309

**Growth, Maturation, and Aging: Development
Continues** 309

Infancy and Childhood: Growing Up Fast 309
Puberty: Sexual Maturity 311
Adulthood: The Longest Stage in the Life Cycle 311
What Causes Aging? 311
Aging Gracefully 312

FEATURES
■ Fertilization in a Laboratory
Dish 290
■ Biology: A Human Endeavor
BOX 14.1: Carl Djerassi:
Population Pessimism,
Contraceptive Concerns 301
■ Personal Impact
BOX 14.2: Sexually Transmitted
Disease: A Growing Concern .. 305
■ Connections 313
Highlights in Review
Key Terms
Study Questions
For Further Reading

PART THREE ■ Life's Variety

CHAPTER 15 ■ Life's Origins and Diversity on
Our Planet 316

Earth As a Stage for Life 318

How the Earth Formed: From Big Bang to Big Rock 318
A Black and Blue World 318
Earth's Advantageous Place in the Sun 319
Ancient Earth and the Raw Materials of Life 319

The Unseen Drama: From Molecules to Cells 320

Molecules Join into Long Chains 321
Chains Copy Themselves 321
Molecular Interactions Take Place 321
Cell-Like Compartments Take Shape 322
Coordinated Cell-Like Activities Emerge 322

Earth and Life Evolve Together 322

Early Life Forms Evolve and Change the Earth 322
The Earth Evolves and Alters Life 326

**The Science of Taxonomy: Cataloging
Life's Diversity** 328

FEATURES

■ Methane Generators: Relatives
  of the Earliest Microbes? 316
■ How Do We Know?
  BOX 15.1: The Search
  Continues 325
■ Focus on the Environment
  BOX 15.2: How Many Species
  Inhabit the Earth? 333
■ Connections 332
  Highlights in Review
  Key Terms
  Study Questions
  For Further Reading

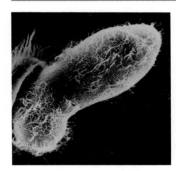

CHAPTER 16 ■ Life As a Single Cell 336

The Lives of Prokaryotic Cells 338

Prokaryotic Cell Structure 338
Nutrition in Prokaryotes 338
Prokaryotic Reproduction 340
Prokaryotic Behavior 340
Types of Prokaryotes 340
The Importance of Prokaryotic Cells 344

Viruses and Other Noncellular Agents of Disease 344

The Protists: Single-Celled Eukaryotes 345

General Characteristics 346
Protozoa: The Animal-Like Protists 346
Carbohydrate Producers: The Plantlike Protists 350
The Funguslike Protists 352
Evolutionary Relationships Among the Protists 353

FEATURES

■ The Complex Life of a
  Microbial Hunter 336
■ Personal Impact
  BOX 16.1: Terminating the
  Terminator 347
■ Personal Impact
  BOX 16.2: Mad Over Mysterious
  Fibrils 351
■ Connections 354
  Highlights in Review
  Key Terms
  Study Questions
  For Further Reading

CHAPTER 17 ■ Plants and Fungi: Decomposers
and Producers 356

**An Overview of the Kingdom Fungi: The Great
Decomposers** 358

The Fungal Body: A Matter of Mycelia 358
Fungal Reproduction: The General Case 358
Fungal Interactions with Plants 359

**A Survey of Saprobes and Parasites: The Major
Fungal Groups** 360

Zygomycota: Bread Molds and Other Zygote Fungi 360
Ascomycota: Mildews and Other Sac Fungi 360
Basidiomycota: Mushrooms and Other Club Fungi 362
Deuteromycota, or Fungi Imperfecti 363

FEATURES

■ Redwoods and Root Fungi 356
■ Focus on the Environment
  BOX 17.1: An Antipollution
  Fungus 367
■ Connections 376
  Highlights in Review
  Key Terms
  Study Questions
  For Further Reading

The Plant Kingdom: An Overview 363

Plant Life Cycles 363
Trends in Plant Evolution: A Changing Planet and the
Challenges of Life on Land 363

Algae: Ancestral Plants That Remained Aquatic 364

Red Algae: Deepest-Dwelling Plants 365
Brown Algae: Giants of the Algal World 366
Green Algae: Ancestors of the Land Plants 366

Simple Land Plants: Still Tied to Water 368

The Bryophytes: Earliest Plant Pioneers 368
Horsetails and Ferns: The Seedless Vascular Plants 369

Gymnosperms: Conquerors of Dry Land 371

Cycads: Cone-Bearing Relics 371
Ginkgos and Gnetophytes: Odd Oldsters 372
Conifers: The Familiar Evergreens 372

Flowering Plants: A Modern Success Story 373

Flowering Plant Reproduction: A Key to Their Success 374
Flowering Plants and Pollinators: A Coevolution 375

CHAPTER 18 ■ Invertebrate Animals: The Quiet
Majority 378

**Animals and Evolution: An Overview of the
Animal Kingdom** 380

Sponges: The Simplest Animals 382

The Sponge's Body Plan 382
Sponge Activities and Reproduction 382

Cnidarians: The Radial Animals 383

"Vases" and "Umbrellas": Cnidarian Body Plans 383
Cnidarians Have Complex Reproductive Cycles 384

**Flatworms: The Beginning of Bilateral Symmetry
and Cephalization** 385

Bilateral Symmetry and Cephalization 386
Distinct Tissues, Organs, and Organ Systems 386
Flatworm Life-Styles 387

Roundworms: Advances in Digestion 387

Roundworm Advances: A Fluid-Filled Body Cavity and
a Two-Ended Gut 388
Impact of Roundworms on the Environment 388

Two Evolutionary Lines of Animals 389

Mollusks: Soft-Bodied Animals of Water and Land 389

Molluscan Body Plan 389
Molluscan Advances 390
Some Classes of Mollusks 390

**Annelid Worms: Segmentation and a Closed
Circulatory System** 391

Annelids Show All Five Evolutionary Trends 392
More Annelid Characteristics 392

Arthropods: The Joint-Legged Majority 393

The Arthropod Exoskeleton 394

FEATURES
■ Self-Protecting Termites 378
■ Connections 399
   Highlights in Review
   Key Terms
   Study Questions
   For Further Reading

Specialized Arthropod Segments 394
Arthropod Respiratory Advances 395
Acute Senses 395
Important Arthropod Classes 395

Echinoderms: The First Endoskeletons 398

CHAPTER 19 ▪ The Chordates: Vertebrates and
Their Relatives 402

**The Chordates, Including the Simple Tunicates
and Lancelets** 404

Tunicates 406
Lancelets 406

Fishes: The Earliest Vertebrates 407

Advances During Fish Evolution 408
Modern Fishes 408

Amphibians: First Vertebrates to Live on Land 410

Reptiles: Conquerors of the Continents 411

Birds: Airborne Vertebrates 413

Mammals: Rulers of the Cenozoic 415

Temperature Regulation 417
Specialized Limbs and Teeth 417
Parental Care 417
Highly Developed Nervous Systems and Senses 417

**Evolution of the Primates: Our Own
Taxonomic Order** 418

The Primate Family Tree 418
Primate Characteristics 421
Primate Evolution 422

The Rise of *Homo sapiens* 423

Homo habilis 423
Homo erectus 423
Homo sapiens 424

FEATURES
▪ Bats: Abundant Vertebrates
with a Bad Reputation 402
▪ Focus on the Environment
BOX 19.1: The Mystery of the
Disappearing Dinosaurs 414
▪ Connections 425
Highlights in Review
Key Terms
Study Questions
For Further Reading

PART FOUR ▪ How Animals Survive

CHAPTER 20 ▪ An Introduction to How
Animals Function 430

**Staying Alive: Problems, Solutions, and the Role of
Physical Size** 432

Solutions to the Central Problem: Homeostasis 432
Body Organization: A Hierarchy of Tissues, Organs, and
Organ Systems 433
The Four Types of Tissues 435
Large Animals Face Additional Problems 435

**Animal Adaptations: Form and Function Suit
the Environment** 437

**Homeostasis: Keeping the Cellular Environment
Constant** 438

Strategies for Homeostasis: Feedback Loops 438
Temperature Regulation: Homeostasis and Evolution 441

FEATURES
▪ A Whale of a Survival
Problem 430
▪ Personal Impact
BOX 20.1: The Case of Martha's
Liver: A Blood Test for
Homeostasis 440
▪ Personal Impact
BOX 20.2: Fever and the Body's
Natural Thermostat 443
▪ Connections 442
Highlights in Review
Key Terms
Study Questions
For Further Reading

CHAPTER 21 ■ Circulation: Transporting Gases
and Materials 446

**Blood: A Multipurpose Liquid Tissue for
Internal Transport** 448

Blood: Liquid and Solids 448
Red Blood Cells and Oxygen Transport 448
White Blood Cells: Defense of the Body 449
Platelets: Plugging Leaks in the System 450

**Circulatory Systems: Strategies for
Material Transport** 450

Open Circulatory Systems 450
Closed Circulatory Systems 451
Blood Circulation 451

**Circulation in Humans and Other Mammals:
The Life-Sustaining Double Loop** 452

Blood Vessels: The Vascular Network 452
The Tireless Heart 454
How the Heart Beats 455
Blood Pressure: The Force Behind Blood Flow 456
Shunting Blood Where It Is Needed: Vasoconstriction
and Vasodilation 458
How Bleeding Stops: The "Clotting Cascade" 459

The Lymphatic System: The Second Fluid Highway 461

FEATURES
■ The Improbable Giraffe 446
■ Personal Impact
BOX 21.1: Cholesterol and Heart
Disease 457
■ Connections 461
Highlights in Review
Key Terms
Study Questions
For Further Reading

CHAPTER 22 ■ The Immune System and
the Body's Defenses 464

**The Body's Defenses: Major Players,
Major Activities** 466

Immediate Nonspecific Protection 466
Specific Immunity Against Specific Targets 466

Antibodies: Defense Molecules 469

Antigen-Binding Sites: Specificity and Diversity 469
How Antibodies Trigger Elimination 470

B Cells: Mobilization and Memory 471

The B-Cell in Action 471
Antibody Synthesis: Reshuffling a "Small" Deck
of Genes 472

T Cells: Direct Combat and Regulation 474

How T Cells Work 474
Immune Regulation: A Balancing Act 476

**Medical Manipulations: Short- and
Long-Term Protection** 477

Passive Immunity: Short-Term Protection by
Borrowed Antibodies 477
Active Immunity: Prevention by Altered Antigen 479

FEATURES
■ AIDS, The New Plague 464
■ How Do We Know?
BOX 22.1: Monoclonal
Antibodies: Tools for Medicine
and Research 470
■ Personal Impact
BOX 22.2: Anatomy of an
Allergy 473
■ Personal Impact
BOX 22.3: Pregnancy, Tolerance,
and Rh Disease 478
■ Connections 480
Highlights in Review
Key Terms
Study Questions
For Further Reading

CHAPTER 23 ■ Respiration: Gas Exchange in Animals 482

**Gas Exchange in Animals: Life-Supporting Oxygen
for Every Cell** 484

Diffusion: The Mechanism of Gas Exchange 484
Extracting Oxygen from Water: The Efficient Gill 484
Adaptations for Respiration in a Dry Environment:
Tracheae and Lungs 486

Respiration in Humans and Other Mammals 487

Respiratory Plumbing: Passageways for Air Flow 487
Ventilation: Moving Air Into and Out of Healthy Lungs 489
Control of Ventilation by the Brain 490

Mechanisms of Gas Exchange 490

Partial Pressure and Diffusion 491
How Hemoglobin Transports Oxygen 491
A Special Mechanism That Unloads Oxygen to
Active Cells 492

FEATURES
■ Antarctic Deep Diver 482
■ Connections 492
  Highlights in Review
  Key Terms
  Study Questions
  For Further Reading

CHAPTER 24 ■ Animal Nutrition and Digestion: Energy
and Materials for Every Cell 494

Nutrients: Energy and Materials That Sustain Life 496

Carbohydrates: Carbon and Energy from Sugars
and Starches 496
Lipids: Highly Compact Energy Storage Nutrients 497
Proteins: Basic to the Structure and Function of Cells 497
Vitamins and Minerals: Nutrients of Great Importance 498
Food as Fuel: Calories Count 502

**Digestion: Animal Strategies for Consuming and
Using Food** 504

The Mechanisms of Extracellular Digestion 505
The Alimentary Canal: Food-Processing Pipeline 506

The Human Digestive System 507

The Mouth, Pharynx, and Esophagus: Mechanical
Breakdown of Food 507
The Stomach: Food Storage and the Start of
Chemical Breakdown 508
The Small Intestine and Accessory Organs: Digestion
Ends, Absorption Begins 509
The Large Intestine: Site of Water Absorption 511

Coordination of Digestion: A Question of Timing 511

FEATURES
■ Up a Tree 494
■ Personal Impact
  BOX 24.1: Fat Cells, Set Point, and
  Weight Loss 503
■ Connections 513
  Highlights in Review
  Key Terms
  Study Questions
  For Further Reading

CHAPTER 25 ■ Excretion and the Balancing of Water
and Salt 516

Ridding the Body of Nitrogenous Wastes 518

Three Forms of Waste in Different Kinds of Animals 518

**The Kidney: Master Organ of Waste Removal,
Water Recycling, and Salt Balance** 520

The Kidneys and Other Organs of the Human
Excretory System 521
The Nephron: Working Unit of the Kidney 522
The Nephrons at Work 524
Kidney Function: A Quick Review 526
Kidneys Adapted to Their Environments 526

**Regulating the Body's Water Content: Thirst
and Hormones** 526

Thirst Regulates Drinking and Helps Maintain
Water Balance 527
Hormones Acting on Nephrons Control the Excretion
of Water 527
The Kidneys' Role in Heart Disease 528

**Strategies for Survival: How Animals Balance Salt
and Water** 529

Osmoregulation in Terrestrial Animals 529
Osmoregulation in Aquatic Animals 530

FEATURES
■ Kangaroo Rats and Capable
Kidneys 516
■ How Do We Know?
BOX 25.1: When the Kidney
Fails 529
■ Connections 532
Highlights in Review
Key Terms
Study Questions
For Further Reading

CHAPTER 26 ■ Hormones and Other Molecular
Messengers 534

Molecular Messengers: An Overview 536

Types of Molecular Messengers 537
Molecular Messengers and Receptors 538

Hormones and the Mammalian Endocrine System 539

Hormone Structure and the Endocrine System 539
Pituitary and Hypothalamus: Controlling the Controllers 540
The Thyroid and Parathyroid Glands: Regulators
of Metabolism 544
The Adrenals: The Stress Glands 544

Hormones, Homeostasis, and Physiological Change 546

Hormones and Homeostasis: Keeping Conditions
Constant 546
Hormones and Cyclic Physiological Changes 547
Hormones and Permanent Developmental Change 548

**Hormones and Evolution: Invariant Agents
of Change** 549

FEATURES
■ The Worm Turns 534
■ How Do We Know?
BOX 26.1: Female Hyenas and
Male Hormones 547
■ Personal Impact
BOX 26.2: Steroid Hormones
and the Instant Physique 549
■ Connections 550
Highlights in Review
Key Terms
Study Questions
For Further Reading

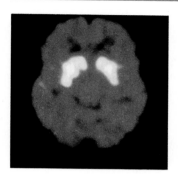

CHAPTER 27 ▪ How Nerve Cells Control Behavior 552

How Nerve Cell Structure Facilitates Communication ..554

Anatomy of a Nerve Cell 554

Electrifying Action: The Nerve Impulse 554

A Nerve Cell at Rest 555
A Neuron's Resting Potential 555
The Action Potential: A Nerve Impulse 557
Propagating the Nerve Impulse 557

How Neurons Communicate Across Synapses 559

Electrical Junctions: Rapid Relayers 560
Chemical Synapses: Communication Across a Cleft 560

How Networks of Neurons Control Behavior 564

Learning Results from Changes at Synapses 564

FEATURES

▪ Cocaine: Highest Highs and
  Lowest Lows 552
▪ Connections 566
  Highlights in Review
  Key Terms
  Study Questions
  For Further Reading

CHAPTER 28 ▪ The Senses and the Brain 568

Window on the World: Sense Organs 570

The Ear: The Body's Most Complex Mechanical Device 570
The Eye: An Outpost of the Brain 572
Taste and Smell: Our Chemical Senses at Work 574

**The Nervous System: From Nerve Net to the Human
Brain** 576

Trends in the Evolution of the Nervous System 576

Organization of the Vertebrate Nervous System 576

The Peripheral Nervous System: The Neural Actors 578

Sensory Neurons 578
Motor Neurons 579

**The Central Nervous System: The Information
Processor** 580

The Spinal Cord: A Neural Highway 580
The Brain: The Ultimate Processor 580
The Brain Stem 580
The Cerebellum: Muscle Coordinator 581
The Cerebrum: Seat of Perception, Thought, Humanness 581

FEATURES

▪ Keen Senses of a Night
  Hunter 568
▪ Personal Impact
  BOX 28.1: Alzheimer's Disease:
  Replacement Parts for
  Damaged Brains 585
▪ Connections 586
  Highlights in Review
  Key Terms
  Study Questions
  For Further Reading

CHAPTER 29 ▪ The Dynamic Animal: The Body
in Motion 588

**The Skeleton: A Scaffold for Support
and Movement** 590

Water as a Skeletal Support 590
Braced Framework Skeletons 590

Muscles: Motors of the Body 594

Protein Filaments: The Muscular Motor 594
Membrane System: The Ignition 596
ATP: The Fuel 599

FEATURES

▪ Battle for an Equine Harem .. 588
▪ Personal Impact
  BOX 29.1: Low Back Pain 597
▪ Connections 602
  Highlights in Review
  Key Terms
  Study Questions
  For Further Reading

Exercise Physiology and Survival 600

Escape or Skirmish: A Survival Response 600
How Athletic Training Alters Physiology 601
Exercise and Heart Disease 602

PART FIVE ▪ How Plants Survive

CHAPTER 30 ▪ Plant Architecture and Function 606

The Plant Body: Plant Tissues and Growth Patterns.....608

The Plant's Main Axis: Root and Shoot 608
Tissue Systems and Tissue Types 608
Open Growth: A Plant's Pattern of Perpetual Growth
and Development 613

**Flowers, Fruits, Seeds, and Plant Embryos:
The Architecture of Continuity** 614

Flowers: Sex Organs from Modified Leaves 615
Pollination, Fertilization, and Seed Formation 615
Germination: The New Plant Emerges 617
Development from Seedling to Pear Tree 617

Structure and Development of Roots 618

The Root: From Tip to Base 618
The Root: From Outside to Inside 619

Structure and Development of the Shoot 622

Stem Structures and Primary Growth 623
Stem Structures and Secondary Growth 623

The Structure and Development of Leaves 625

Leaf Blade and Petiole 627
Anatomy of a Leaf...627

FEATURES
▪ A Blooming Pear Tree 606
▪ How Do We Know?
  BOX 30.1: The Answer Was
  Blowing in the Wind 616
▪ Personal Impact
  BOX 30.2: Exotic Plants for a
  Hungry Planet 620
▪ Connections 628
  Highlights in Review
  Key Terms
  Study Questions
  For Further Reading

CHAPTER 31 ▪ Regulators of Plant Growth
and Development 630

Plant Hormones: Five Major Kinds 632

Gibberellins 632
Auxins 632
Cytokinins 633
Abscisic Acid 633
Ethylene 633

How Hormones Control Germination 634

Environmental Cues, Hormones, and Germination 634
How Plant Biologists Learned the Effects of Hormones
on Germination 634

Regulation of Plant Growth and Development 635

How Environmental Cues Can Influence
a Plant's Orientation 635
Plant Movements Other Than Tropisms 637
How the Environment Influences a Plant's Shape 637

Control of Flowering and Fruit Formation 639

Temperature and Flowering 639

FEATURES
▪ Foolish Seedling Disease 630
▪ Connections 644
  Highlights in Review
  Key Terms
  Study Questions
  For Further Reading

Light and Flowering 639
The Role of Hormones in Flowering 640
Triggers to Fruit Development 641

**How Plants Age and How They Prepare
for Winter** 641

Senescence and Abscission 642
Dormancy: The Plant at Rest 642

**Plant Protection: Defensive Responses
to External Threats** 643

Chemical Protection 643
Walling Off Injured Areas 644

CHAPTER 32 ▪ The Dynamic Plant: Transporting Water
and Nutrients 646

How Plants Take Up Water and Restrict Its Loss 648

How Roots Draw Water from the Soil 648
Water Transport: Root to Leaf 648
Root Pressure, Xylem Pumps, and Capillary Action 649
Transpiration: The Life-Giving Chain of Water 650
Water Stress: A Break in the Chain 650
Stomata and the Regulation of Water Loss 651

How Plants Absorb Needed Mineral Nutrients 651

What Nutrients Do Plants Need? 652
Soil: The Primary Source of Minerals 654
How Nutrients Enter a Plant 655

Moving Organic Molecules Throughout the Plant 656

Translocation: How Sugar Moves from Source to Sink 656
Transport of Inorganic Substances in Xylem and Phloem 657

Engineering Useful Plants 658

New Techniques of Artificial Selection 659
Genetic Engineering in Plants 660

FEATURES
▪ Tomatoes: A Case Study for
Plant Transport 646
▪ Personal Impact
BOX 32.1: Retooling the
Tasteless Tomato 660
▪ Connections 661
Highlights in Review
Key Terms
Study Questions
For Further Reading

PART SIX ▪ Interactions: Organisms and Environment

CHAPTER 33 ▪ The Genetic Basis for Evolution 664

Genetic Variation: The Raw Material of Evolution 666

What Is the Extent of Genetic Variation? 666
What Are the Sources of Genetic Variation? 667
How Is Genetic Variation Maintained? 668

The Agents of Evolution 669

Mutation As an Agent of Evolution 669
Migration Alters Allele Frequencies 669
Chance Changes in Small Populations 669
Nonrandom Mating and Evolution 671
Selection 672
Which Agent of Evolution Is the Most Important? 672

Natural Selection in Action 673

Sickle Cell Anemia and Natural Selection 674
Some Modes of Selection 674

FEATURES
▪ Cheetahs: Sprinting Toward
Extinction 664
▪ How Do We Know?
BOX 33.1: The Hardy-Weinberg
Principle: How Do We Know a
Population Is Evolving? 670
▪ Connections 688
Highlights in Review
Key Terms
Study Questions
For Further Reading

How New Species Arise 676
What Is a Species? 676
Reproductive Isolating Mechanisms 677
The Origin of New Species 678

Physical Evidence for Evolution 679
Life's History in Stone: Evidence from the
Fossil Record 679
Molecular Evidence for Evolution 680
Evidence from Anatomy: Homology, Vestigial Organs,
and Comparative Embryology 681
Evidence from Ecology: Convergent Evolution and
Biogeography 683

Trends in Macroevolution 683
Patterns of Descent 683
The Tempo of Evolution 685
Mechanisms of Macroevolution 686

CHAPTER 34 ■ Population Ecology: Patterns in Space
and Time 690

The Science of Ecology: Levels of Interaction 692

Distribution Patterns: Where Do Populations Live? 693

Limits to Global Distribution 693
Local Patterns of Distribution 693

Constraints on Population Size 695

Factors Affecting Population Size 695
Limited Resources Limit Exponential Growth 698
Logistic Growth and Strategies for Survival 702

**The Past, Present, and Future of the
Human Population** 703

Trends in Human Population Growth 703
Underlying Causes of Change in Human Population Size 703
Growth Rates and Age Structure 706
What Causes Birthrates to Decline? 706
Population of the Future 707

FEATURES
■ The Rise and Fall of a Desert
Population 690
■ Focus on the Environment
BOX 34.1: Amphibians: Missing
in Action 700
■ Connections 707
Highlights in Review
Key Terms
Study Questions
For Further Reading

CHAPTER 35 ■ The Ecology of Communities:
Populations Interacting 710

**Habitat and Niche: Where Organisms Reside and
How They Live** 712

The Many Ways Species Interact 713

Competition Between Species for Limited Resources 713

How Species Might Compete in Communities 713
Competition in Real Communities: Natural Causes of
Competitive Exclusion 714
Competition Can Alter a Species' Realized Niche 715

The Hunter and The Hunted: Predation 716

Populations of Predator and Prey 716
The Coevolutionary Race Between Predator and Prey 718
Parasites: The Intimate Predators 721

**Sharing and Teamwork: Commensalism
and Mutualism** 722

FEATURES
■ The Intertwined Lives of
Flower and Moth 710
■ Focus on the Environment
BOX 35.1: Integrated Pest
Management: Community
Ecology Applied to
Agriculture 720
■ Focus on the Environment
BOX 35.2: Community Ecology
of Islands and the Design of
Nature Preserves 727
■ Connections 728
Highlights in Review
Key Terms
Study Questions
For Further Reading

Commensalism 723
Mutualism 723

**How Communities Are Organized in Time
and Space** 724

Communities Change Over Time 724
Trends in Species Diversity in Communities 725
Species Diversity, Community Stability, and Disturbances 726

CHAPTER 36 ▪ Ecosystems: Webs of Life and the
Physical World 730

**Pathways for Energy and Materials: Who Eats
Whom in Nature?** 732

Feeding Levels: Strategies for Obtaining Energy 732
Feeding Patterns in Nature 733

How Energy Flows Through Ecosystems 733

Energy Budget for an Ecosystem 733
Energy and Trophic Levels: Ecological Pyramids 736

How Materials Cycle Through Ecosystems 738

The Water Cycle: Driven by Solar Power 739
The Nitrogen Cycle Depends on Nitrogen-Fixing Bacteria..... 739
Phosphorus Cycles Locally and in Geological Time 741
The Carbon Cycle: Coupled to the Flow of Energy 743

**How Human Intervention Alters Ecosystem
Function** 743

Global Warming 743
When It Rains, It Pours Sulfuric Acid 746
Cycle and Recycle: The Sustainable Economy 746
Nuclear Winter: A Catastrophic Ecological Threat 747

FEATURES
▪ A Voyage of Discovery 730
▪ How Do We Know?
  BOX 36.1: Lemmings in the
  Tundra: Case Study of
  Ecosystem Dynamics 742
▪ Focus on the Environment
  BOX 36.2: Saving the
  Environment: Personal
  Solutions 747
▪ Connections 748
  Highlights in Review
  Key Terms
  Study Questions
  For Further Reading

CHAPTER 37 ▪ The Biosphere: Earth's Thin
Film of Life 750

What Generates the Earth's Climatic Regions? 752

A Round, Tilted Earth Heats Unevenly 752
The Formation of Rain 752
Air and Water Currents: The Genesis of Wind
and Weather 753

Biomes: Life on the Land 754

Tropical Rain Forests 756
Tropical Savannas 757
Deserts 758
Temperate Grasslands 759
Chaparral 760
Temperate Forests 760
Coniferous Forests 761
Tundra 761
Polar Cap Terrain 762
Life Zones in the Mountains 762

FEATURES
▪ The Cathedral Forest 750
▪ Focus on the Environment
  BOX 37.1: Attack on the
  Ozone 768
▪ Focus on the Environment
  BOX 37.2: Sustainability: A New
  World Order 769
▪ Connections 769
  Highlights in Review
  Key Terms
  Study Questions
  For Further Reading

Life in the Water 763

Properties of Water 763
Freshwater Communities 763
Saltwater Communities 765

Change in the Biosphere 766

PART SEVEN ▪ Behavior and the Future

CHAPTER 38 ▪ Animal Behavior: Adaptations
for Survival 774

**Genes Direct Development and Behavioral
Capabilities** 776

**Experiences in the Environment Can Alter
Behavior** 777

Innate Behaviors: Automatic Responses to the Environment .. 777
Behaviors with Limited Flexibility 778
Imprinting: Learning with a Time Limit 779
Learning: Behavior That Changes with Experience 780

How Natural Selection Shapes Behaviors 782

Locating and Defending a Home Territory 782
Feeding Behavior: Finding Food While Avoiding
Predators 784
Reproductive Behavior: The Central Focus of
Natural Selection 785
Communication: Messages That Enhance Survival 787

The Evolution of Social Behavior 788

Social Living: Advantages and Disadvantages 789
The Evolution of Altruistic Behavior 789
Sociobiology and Human Behavior 790

FEATURES
▪ Menace of the Broken Shell .. 774
▪ Connections 791
Highlights in Review
Key Terms
Study Questions
For Further Reading

Appendix A-1
Glossary G-1
Credits and Acknowledgments C-1
Index I-1

PREFACE

Modern biology is often one of the most popular undergraduate science courses at colleges and universities. Many students enroll because they'd like to learn more about human physiology, genetics, and the environment. Some sign up because science is required and they hope that biology will be easier than chemistry or physics. And a handful mainly need a M–W–F at 10:40.

Reasons aside, this will be the first and last exposure many nonmajors will have to a science course during their college years, and it presents their best opportunity to become biologically literate citizens who can make wise decisions about nutrition, exercise, health care, drugs, alcohol, smoking, sexually transmitted diseases, consumer products, habits that impact the environment, and important issues before the electorate.

While introductory biology has obvious relevance to a student's life, the course can nevertheless be challenging—for some students, even formidable—because of the sheer number of facts and terms that biologists use in their work and that students have traditionally been required to learn. The well-documented trend toward science illiteracy among Americans is partially caused by the burgeoning of facts and terms in all scientific fields, biology included. But the expansion of science is just one factor—and considering the higher levels of science literacy in a dozen other industrialized nations, perhaps a minor one. More to the point, say many science educators, is the way we have customarily presented science from elementary school through graduate school.

The National Science Foundation has established a grant program to encourage new approaches to the teaching of undergraduate science courses. The collective innovations proposed by professors at dozens of colleges and universities represent an emerging consensus that emphasizes:

- Concepts over details
- Issues and student-relevant topics over traditional vocabulary and facts
- Hands-on, self-generated laboratory projects over teacher demonstrations and "cookbook" exercises
- Writing and speaking skills over memorization and standard exams
- Cooperative learning by small groups in addition to attendance at large lectures

Overall, life science professors want their students to experience the excitement and to see the applications of modern biology; to carry away an understanding of basic concepts, not a memory full of transitory facts; to appreciate science as a way of knowing about the world around them; and to apply biological concepts to the health, reproductive, and environmental choices they will inevitably face.

Features from the First Edition

Our approach to the first edition of *The Nature of Life* reflected a number of these innovations and was enthusiasti-

cally received in many colleges and universities. To motivate and excite nonmajors, we generated 38 stories, one per chapter, around which to pose and answer central questions about the natural world and to organize basic biological concepts. We mapped out three bookwide unifying themes to help orient readers:

1. Living things take in *energy,* needed to maintain their internal order and organization.
2. Living things undergo *reproduction,* enabling the species to continue after the individual ceases to exist.
3. Specialized means of acquiring energy and characteristic patterns of reproduction arise by *evolution,* allowing living organisms to adapt to changing environments.

The first edition included chapterwide themes, as well, to help establish the relationship of parts to the whole and to guide students to what is most significant, and why. What's more, we created an art program full of orienting icons, unique process diagrams, and colorful photos, and closely coordinated them to the text so that readers could verbalize and visualize biological structures and activities simultaneously. We incorporated current issues and topics of student relevance throughout the chapters and the more than 50 boxed essays. Finally, we added several special study features—underlined take-home messages, integrating end-of-chapter reflections called Connections, Highlights in Review, Key Terms, Study Questions, and Further Readings—to help the student learn biology most effectively. Our first edition provided a guided tour of the features in a typical chapter, and an updated version appears in this new edition, beginning on page xxx.

As in the first edition of *The Nature of Life,* this new revision takes a hierarchical approach to the study of biology. An introductory chapter discusses the main themes of the book as well as the process of science. In Part One, we consider how molecules (Chapter 2), cells (Chapter 3), and cellular activities (Chapter 4) provide the common groundwork for life. We end Part One with a clear analysis of how cells obtain and use energy (Chapters 5 and 6), the book's first theme.

From this cellular foundation, we move on to the principles of reproduction, the book's second theme, and the general subject of Part Two. Chapters 7 through 10 discuss how cells and organisms pass on to their offspring the hereditary units that cause "like to beget like." A short, very up-to-date chapter on the exciting world of biotechnology and recombinant DNA techniques (Chapter 11) leads in to the fascinating subject of human genetics (Chapter 12). This part of the book ends with an analysis of how egg cells decode instructions stored in DNA to build a fish, fly, frog, or person (Chapters 13 and 14).

Equipped with a thorough knowledge of the cellular and genetic features that unify life, we survey the diverse range of life forms in Part Three. We take an evolutionary approach (the book's third theme) to the questions of life's origins (Chapter 15). Then we examine the characteristics of organisms from each kingdom of life (Chapters 16 through 19), ending with an in-depth discussion of the human species' own evolutionary descent.

Parts Four and Five investigate how plants and animals maintain their bodies against the inevitable disorganization that occurs over time (a reprise of our first theme). Chapter 20 orients students by presenting the themes that will recur throughout the study of anatomy and physiology. Chapters 21 through 28 analyze individual physiological systems in animals and include dozens of examples that we hope will captivate readers and allow them to understand the biological bases for their own body functions. We include a complete discussion of the immune system (Chapter 22), an in-depth treatment of the nervous system, brain, and behavior (Chapters 27 and 28), and a unique chapter that focuses on exercise physiology (Chapter 29) and that shows how body systems work together as a dynamic whole. This chapter promises to have great appeal for physically active college students.

Part Five builds on the discussion of plant diversity from Chapter 17 with three more plant chapters: one on plant structure (Chapter 30), one on the regulation of plant growth (Chapter 31), and one that deals with how plants function (Chapter 32). These chapters incorporate the many new applications of genetic engineering to plant science, and they emphasize the importance of plants to students' everyday lives.

With Part Six we return to our evolutionary theme, devoting Chapter 33 to the science of evolution, and Chapters 34 through 37 to ecology, how organisms interact with each other and the world around them. We give special emphasis to the ecological and environmental issues that affect our current quality of life and that of future generations.

The book ends with the engrossing subject of behavior (Part Seven, Chapter 38), viewed from evolutionary and ecological perspectives. We believe this is a fitting conclusion because human behavior will shape the world of the future—a world our students must prepare to lead as professionals and to help preserve as responsible citizens.

New to This Edition
Although we retain this basic organization in our current revision of *The Nature of Life,* we made numerous changes, reflecting both our reviewers' helpful advice and new ideas. Our goal was to continue motivating and exciting introductory biology students in ways that support both traditional and innovative courses, but to do so in an even more effective manner:

- We cut detail in many discussions, simplifying the overall presentation and strengthening the concepts and take-home messages.
- We updated the science where appropriate in the text and boxed essays.

- We significantly increased our coverage of environmental science and added the preservation of the environment as a new bookwide theme.
- We enlarged and strengthened our section on the physical evidence for evolution.
- We introduced more human relevance, new societal issues, and new opening examples in many chapters.
- We streamlined dozens of diagrams to improve their readability, and we replaced or enlarged many photos to integrate the visual with the verbal more successfully.
- We sharpened our presentation of the scientific process.
- We enhanced the end-of-chapter questions and bibliographies.
- We redesigned the book's graphic elements to make the book easier to read, clearer, and more inviting.

We hope that the second edition of *The Nature of Life* will be a useful educational tool. We have tried to preserve the best features of the first edition while improving upon them in innovative ways that will intrigue nonmajors and help them learn the basics of biological science and then apply that knowledge to their own lives.

This second edition is the designated textbook to accompany a planned college-level biology telecourse and an eight-part prime-time series coming to public television, produced by WGBH Boston.

SUPPLEMENTARY MATERIALS

A comprehensive and completely integrated package of supplementary materials accompanies *The Nature of Life,* Second Edition.

- ***Instructor's Manual and Resource Guide***
  Dennis Todd, University of Oregon
- ***Test Bank***
  Dennis Todd, University of Oregon
- ***Critical Thinking Workbook to Accompany* The Nature of Life**
  Gail Patt, Boston University
- ***Hands-On Biology*** (a laboratory manual for introductory biology) and ***Preparator's Guide to Accompany* Hands-On Biology**
  Theodore Taigen, University of Connecticut, Storrs
  Thomas Terry, University of Connecticut, Storrs
  David Wagner, University of Connecticut, Storrs
  Eileen Jokinen, University of Connecticut, Storrs
  Andris Indars, University of Connecticut, Storrs
- Computerized Instructor's Manual (available in IBM, Macintosh, Apple)
- Computerized Test Bank (available in IBM, Macintosh, Apple)
- ***BioPartner*** (Computerized Study Guide; available in IBM, Macintosh)
- Videos

- Biology Slides and Acetate Package
- Videodisk
- HyperMedia Software

For further information regarding the supplements available, please contact your local McGraw-Hill representative.

ACKNOWLEDGMENTS

This revision grew out of the original material we presented in the first edition, upon the insightful comments of our many reviewers and marketing consultants, and finally upon our new manuscript and illustrations so skillfully handled by the McGraw-Hill team. We wish, therefore, to thank those central to the inception and development of *The Nature of Life:* our former sponsoring editor, Eirik Børve; our former developmental editor, Ruth Veres; our former developmental art consultant, Peter Veres; our seven original contributors, Charles L. Aker, The National Autonomous University of Nicaragua; Russ Fernald, University of Oregon; Craig Heller, Stanford University; Kent Holsinger, University of Connecticut; V. Pat Lombardi, Stanford University; Christopher Stringer, British Museum, London; and Daniel Udovic, University of Oregon; and our general consultant, the late Howard Schneiderman, whose counsel, support, and enthusiasm were so important over the years. We also extend special thanks to Dennis Todd from the University of Oregon for his design of Study Questions and material in both editions.

We have sought the advice of hundreds of instructors around the country to help us create a textbook that would meet the unique needs of the introductory biology market. Our sincere thanks are extended to the following individuals who responded to our market questionnaires:

Dr. Laura Adamkewicz, George Mason University; *Olukemi Adewusi,* Ferris State University; *Kraig Adler,* Cornell University; *Dr. John U. Aliff,* Glendale Community College; *Joanna T. Ambron,* Queensborough Community College; *Steven Austad,* Harvard University; *Robert J. Baalman,* California State University, Hayward; *Stuart S. Bamforth,* Tulane University; *Sarah F. Barlow,* Middle Tennessee State University; *R. J. Barnett,* California State University, Chico; *Joseph A. Beatty,* Southern Illinois University, Carbondale; *Nancy Benchimol,* Nassau Community College; *Dr. Rolf W. Benseler,* California State University, Hayward; *Gerald Bergtrom,* University of Wisconsin, Milwaukee; *Dr. Dorothy B. Berner,* Temple University; *Dr. A. K. Boateng,* Florida Community College, Jacksonville; *William S. Bradshaw,* Brigham Young University; *Jonathan Brosin,* Sacramento City College; *Howard E. Buhse, Jr.,* University of Illinois, Chicago; *John Burger,* University of New Hampshire; *F. M. Butterworth,* Oakland University; *Guy Cameron,* University of Houston; *Ian M. Campbell,* University of Pittsburgh; *John L. Caruso,* University of Cincinnati; *Brenda Casper,* University of Pennsylvania; *Doug Cheeseman,* De Anza College; *Dr. Gregory Cheplick,* University of Wisconsin; *Dr. Joseph P.*

Chinnici, Virginia Commonwealth University; *Carl F. Chuey,* Youngstown State University; *Dr. Simon Chung,* Northeastern Illinois University; *Norman S. Cohn,* Ohio University; *Paul Colinvaux,* Ohio State University; *Scott L. Collins,* University of Oklahoma; *Dr. August J. Colo,* Middlesex County College; *G. Dennis Cooke,* Kent State University; *Jack D. Cote,* College of Lake County; *Gerald T. Cowley,* University of South Carolina; *Louis Crescitelli,* Bergen Community College; *Orlando Cuellar,* University of Utah; *Thomas Daniel,* University of Washington; *J. Michael DeBow,* San Joaquin Delta College; *Loren Denny,* Southwest Missouri State University; *Ron DePry,* Fresno City College; *Dr. Kathryn Dickson,* California State University, Fullerton; *Patrick J. Doyle,* Middle Tennessee State University; *Dr. David W. Eldridge,* Baylor University; *Lynne Elkin,* California State University, Hayward; *Paul R. Elliott,* Florida State University; *Eldon Enger,* Delta College; *Gauhari Farooka,* University of Nebraska, Omaha; *Marvin Fawley,* North Dakota State University; *Ronald R. Fenstermacher,* Community College of Philadelphia; *Edwin Franks,* Western Illinois University; *C. E. Freeman,* University of Texas, El Paso; *Lawrence D. Friedman,* University of Missouri, St. Louis; *Dr. Ric A. Garcia,* Clemson University; *Wendell Gauger,* University of Nebraska, Lincoln; *Dr. S. M. Gittleson,* Fairleigh Dickinson University; *E. Goudsmit,* Oakland University; *John S. Graham,* Bowling Green State University; *Shirley Graham,* Kent State University; *Thomas Gregg,* Miami University; *Alan Groeger,* Southwest Texas State University; *Thaddeus A. Grudzien,* Oakland University; *James A. Guikema,* Kansas State University; *Robert W. Hamilton,* Loyola University of Chicago; *Richard C. Harrel,* Lamar University; *T. P. Harrison,* Central State University; *Maurice E. Hartman,* Palm Beach Community College; *Dr. Karl H. Hasenstein,* University of Southwestern Louisiana; *Martin A. Hegyi,* Fordham University; *Dr. John J. Heise,* Georgia Institute of Technology; *H. T. Hendrickson,* University of North Carolina, Greensboro; *T. R. Hoage,* Sam Houston State University; *Kurt G. Hofter,* Florida State University; *Dr. Rhodes B. Holliman,* Virginia Polytechnic Institute and State University; *Harry L. Holloway,* University of North Dakota; *E. Bruce Holmes,* Western Illinois University; *Jerry H. Hubschman,* Wright State University; *Hadar Isseroff,* State University of New York College, Buffalo; *Dr. Ira James,* California State University, Long Beach; *Dr. Wilmar B. Jansma,* University of Northern Iowa; *Dr. Margaret Jefferson,* California State University, Los Angeles; *Dr. Ira Jones,* California State University, Long Beach; *Dr. Patricia P. Jones,* Stanford University; *Dr. Craig T. Jordan,* University of Texas, San Antonio; *Maurice C. Kalb,* University of Wisconsin, Whitewater; *Bonnie Kalison,* Mesa College; *Judy Kandel,* California State University, Fullerton; *Arnold Karpoff,* University of Louisville; *L. G. Kavaljian,* California State University, Sacramento; *Donald R. Kirk,* Shasta College; *R. Koide,* Pennsylvania State University; *Mark Konikoff,* University of Southwestern Louisiana; *Barbara S. Lake,* Central Piedmont Community College; *Jim des Lauvérs,* Chaffey College; *Tami*

Levitt-Gilmarr, Pennsylvania State University; *Daniel Linzer*, Northwestern University; *J. R. Loewenberg*, University of Wisconsin, Milwaukee; *Dr. Robert Lonard*, University of Texas, Pan American; *Sharon R. Long*, Stanford University; *Carmita E. Love*, Community College of Philadelphia; *C. E. Ludwig*, California State University, Sacramento; *Dr. Ann S. Lumsden*, Florida State University; *Dr. Bonnie Lustigman*, Montclair State College; *Edward B. Lyke*, California State University, Hayward; *Douglas Lyng*, Indiana University–Purdue University, Ft. Wayne; *George L. Marchin*, Kansas State University; *Philip M. Mathis*, Middle Tennessee State University; *Mrs. Margaret L. May*, Virginia Commonwealth University; *Edward McCrady*, University of North Carolina, Greensboro; *Bruce McCune*, Oregon State University; *Dr. John O. Mecom*, Richland College; *Tekié Mehary*, University of Washington; *Richard L. Miller*, Temple University; *Phyllis Moore*, University of Arkansas, Little Rock; *Carl Moos*, State University of New York, Stony Brook; *Doris Morgan*, Middlesex County College; *Donald B. Morzenti*, Milwaukee Area Technical College; *Steve Murray*, California State University, Fullerton; *Robert Neill*, University of Texas, Arlington; *Paul Nollen*, Western Illinois University; *Kenneth Nuss*, University of Northern Iowa; *William D. O'Dell*, University of Nebraska, Omaha; *Dr. Joyce K. Ono*, California State University, Fullerton; *James T. Oris*, Miami University; *Clark L. Ovrebo*, Central State University, Edmond; *Charles Page*, El Camino College; *Kay Pauling*, Foothill College; *Dr. Chris E. Petersen*, College of DuPage; *Richard Petersen*, Portland State University; *Jeffrey Pommerville*, Glendale Community College; *David I. Rasmussen*, Arizona State University; *Daniel Read*, Central Piedmont Community College; *Dr. Don Reinhardt*, Georgia State University; *Louis Renaud*, Prince George's Community College; *Jackie Reynolds*, Richland College; *Jennifer H. Richards*, Florida International University; *Thomas L. Richards*, California State Polytechnic University; *Tom Rike*, Glendale Community College, California; *C. L. Rockett*, Bowling Green State University; *Hugh Rooney*, J. S. Reynolds Community College; *Wayne C. Rosing*, Middle Tennessee State University; *Frederick C. Ross*, Delta College; *A. H. Rothman*, California State University, Fullerton; *Mary Lou Rottman*, University of Colorado, Denver; *Dr. Donald J. Roufa*, Kansas State University; *Dr. Michael Rourke*, Bakersfield College; *Chester E. Rufh*, Youngstown State University; *Mariette Ruppert*, Clemson University; *Charles L. Rutherford*, Virginia Polytechnic Institute and State University; *Dr. Milton Saier*, University of California, San Diego; *Lisa Sardinia*, San Francisco State University; *A. G. Scarbrough*, Towson State University; *Dan Scheirer*, Northeastern University; *Randall Schietzelf*, Harper College; *Robert W. Schuhmacher*, Kean College of New Jersey; *Joel S. Schwartz*, College of Staten Island; *Roger S. Sharpe*, University of Nebraska, Omaha; *Stanley Shostak*, University of Pittsburgh; *J. Kenneth Shull, Jr.*, Appalachian State University; *C. Steven Sikes*, University of South Alabama; *Christopher C. Smith*, Kansas State University; *John O. Stanton*, Monroe Community College; *D. R. Starr*, Mt. Hood Community College; *Dr. Ruth B. Thomas*, Sam Houston State University;

Nancy C. Tuckman, Loyola University of Chicago; *Dr. Spencer Jay Turkel*, New York Institute of Technology; *William A. Turner*, Wayne State University; *C. L. Tydings*, Youngstown State University; *John Tyson*, Virginia Polytechnic Institute and State University; *Richard R. Vance*, University of California, Los Angeles; *Harry van Keulen*, Cleveland State University; *Roy M. Ventullo*, University of Dayton; *Judith A. Verbeke*, University of Illinois, Chicago; *Dr. Ronald B. Walter*, Southwest Texas State University; *Stephen Watts*, University of Alabama, Birmingham; *Dr. Joel D. Weintraub*, California State University, Fullerton; *Marion R. Wells*, Middle Tennessee State University; *James White*, New York City Technical College; *Joe Whitesell*, University of Arkansas, Little Rock; *Fred Whittaker*, University of Louisville; *Roberta Williams*, University of Nevada, Las Vegas; *Chuck Wimpee*, University of Wisconsin, Milwaukee; *Mala Wingerd*, San Diego State University; *Richard Wise*, Bakersfield College; *Gary Wisehart*, San Diego City College; *Dan Wivagg*, Baylor University; *Richard P. Wurst*, Central Connecticut State University; *Edward K. Yeargers*, Georgia Institute of Technology; *Linda Yasui*, Northern Illinois University.

In addition, many users of the first edition graciously provided us with feedback, which allowed us to focus on areas that could be enhanced or modified. They are:

Dean A. Adkins, Marshall University; *Leslie Drew*, Texas Tech University; *Douglas J. Eder*, Southern Illinois University, Edwardsville; *Richard Haas*, Cal State University, Fresno; *Earl L. Hanebrink*, Arkansas State University; *Dr. John P. Harley*, Eastern Kentucky University; *Marcia Harrison*, Marshall University; *Frank Heppner*, University of Rhode Island; *B. Hunnicutt*, Seminole Community College; *Ursula Jando*, Washburn University of Topeka; *Norma G. Johnson*, University of North Carolina, Chapel Hill; *Clyde Jones*, Texas Tech University; *Thomas L. Keefe*, Eastern Kentucky University; *Robin C. Kennedy*, University of Missouri, Columbia; *Eugene C. Perri*, Bucks County Community College; *Eugene C. Perri*, Bucks County Community College; *Joel B. Piperberg*, Millersville University of Pennsylvania; *William D. Rogers*, Ball State University; *Dr. Fred Schreiber*, California State University, Fresno; *Dr. Jane R. Shoup*, Purdue University, Calumet; *Joseph D. Stogner*, Ferrum College; *Bert Tribbey*, California State University, Fresno; *John Twente*, University of Missouri, Columbia; *Leonard S. Vincent*, Fullerton College; *T. Weaver*, Montana State University; *Kenneth A. Wilson*, California State University, Northridge; *Thomas Wolf*, Washburn University of Topeka; *Paul Wright*, Western Carolina University.

We express our sincere appreciation to reviewers who carefully reviewed the first-edition textbook or drafts of the second-edition manuscript and provided extensive comments and advice. They are:

Aimée H. Bakken, University of Washington; *Jack Bostrack*, University of Wisconsin, River Falls; *Charlotte Clark*, Fullerton College; *David Darda*, Central Washington University; *Kathryn A. Dickson*, California State University, Fullerton;

Thomas Dolan, Butler University; *Helen Dunlap,* Millersville University of Pennsylvania; *Grace Gagliardi,* Bucks County Community College; *Gregory Grove,* Pennsylvania State University; *Madeline M. Hall,* Cleveland State University; *Robert N. Hurst,* Purdue University; *Jerry Kaster,* University of Wisconsin, Milwaukee—Center for Great Lakes Studies; *Robin C. Kennedy,* University of Missouri, Columbia; *Eliot Krause,* Seton Hall University; *Elmo A. Law,* University of Missouri, Kansas City; *James Luken,* Northern Kentucky University; *Gail Patt,* Boston University; *David Polcyn,* California State University, San Bernardino; *Michael Pollock,* Mount Royal College, Canada; *Deborah D. Ross,* Indiana University–Purdue University, Ft. Wayne; *Erik P. Scully,* Towson State University; *Guy Steucek,* Millersville University of Pennsylvania; *Raymond Tampari,* Northern Arizona University; *William A. Turner,* Wayne State University; *Linda Van Thiel,* Wayne State University; *C. David Vanicek,* California State University, Sacramento; *Thomas Wolf,* Washburn University of Topeka.

Finally, we are enormously grateful to the fine professionals at McGraw-Hill who directed our work so smoothly—sponsoring editors Denise Schanck and June Smith and senior associate editor Mary Eshelman—and who polished and produced our manuscript so artfully—Alice Mace Nakanishi, senior editing supervisor; Marian Hartsough, art coordinator; Iris Martinez Kane, Cherie Wetzel, and Arthur Ciccone, art reviewers; Darcy Lanham and Monica Suder, photo researchers; Pattie Myers, assistant production manager; Janet Greenblatt, copyeditor; Sarah Miller, proofreader; Tess Joseph, typist; and Lesley Walsh, office manager.

If this edition of *The Nature of Life* inspires and informs the undergraduate students of the early 1990s, it will be in large measure because the above-mentioned contributed their ideas and energy so generously to help us improve this new version.

John H. Postlethwait and Janet L. Hopson

A Guided Tour to THE NATURE OF LIFE

Central Example

▶ ▶ ▶ ▶ ▶ ▶ ▶ ▶ ▶ ▶ ▶ ▶ ▶ ▶ ▶ ▶ ▶ ▶

The student is led through an intriguing real-world narrative to the main chapter concepts. This popular feature from the first edition has been written in a style that motivates the student to read further.

CHAPTER ▪ 10

How Genes Work: From DNA to RNA to Protein

Cystic Fibrosis: A Case Study in Gene Action

A tiny infant receives a parent's tender kiss on the cheek, and the mother or father receives a piece of disquieting information: The baby tastes so *salty*. Why? This exchange may be the first indication that the newborn has cystic fibrosis, the most common lethal genetic disease among Caucasians. One in every 20 people of northern European ancestry carries one copy of the recessive mutation which leads to the disease, but shows no symptoms. Only a child who inherits one copy from each parent (and is thus a homozygous recessive like a pea plant with a short stem or white flowers) will develop this life-threatening illness.

Cystic fibrosis is essentially a disease of clogged ducts; the recessive gene and the faulty protein it encodes (described shortly) lead to a buildup of sticky mucus in lung passages, pancreas ducts, sweat glands, and sperm ducts. As a result, the individual with cystic fibrosis tends to have difficulty breathing, as well as dangerous bacterial infections in the lungs, stomachaches due to poor absorption of essential fats in the diet, a salty secretion on the skin, and, in adult males, usually sterility.

Right now, doctors must treat the symptoms of cystic fibrosis one by one rather than correcting the genetic defect itself. As we will see later in the chapter, that basic genetic correction may be possible in the not-so-distant future and is one of the most exciting medical prospects for the early 1990s. The current treatments, however, include mainly "percussion sessions" during which a parent thumps the child firmly on the back to dislodge gummy mucus and allow easier breathing for a while (Figure 10.1); a special diet; powerful antibiotics to fight the lung infections; and pills containing certain digestive

FIGURE 10.1 ▪ Manual Therapy for a Child with Cystic Fibrosis. A "percussion session," during which a mother thumps her child's back vigorously, to loosen and help expel the sticky mucus clogging the victim's lungs. This therapy decreases the infections, scarring, and gradual loss of lung function that characterize cystic fibrosis. A mutated gene leads to this devastating condition.

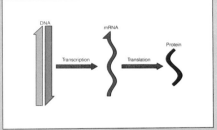

FIGURE 10.2 ▪ **Path of Information Flow in a Cell: DNA to RNA to Protein.** Enzymes transcribe DNA into a messenger RNA molecule; ribosomes translate the mRNA into the polypeptide chain of a protein.

enzymes to replace those blocked up in the pancreas by clogged ducts. Despite these symptomatic treatments, nearly half of the affected individuals fail to survive to age 20. That's why the prospect of gene therapy for cystic fibrosis is so important.

How does a mutation in a single gene cause defects in so many systems (the lungs, pancreas, sweat glands, sperm ducts)? What makes one human gene out of approximately 100,000 so important? And how can physicians combat the effects of the deadly gene more effectively? The answers to these questions are crucial to the health of about 50,000 young people in the United States and are based on the principles of gene action discussed in this chapter.

We know that each gene specifies the synthesis of one polypeptide (see Chapter 9), so it is not surprising that a disease like cystic fibrosis—the result of a single defective gene—results from errors in the synthesis of a single protein. In cystic fibrosis, the protein is called CFTR (for cystic fibrosis transmembrane regulator). This protein is embedded in the cell's plasma membrane and probably transports chloride ions (Cl⁻) out of the cell. Sufferers of cystic fibrosis make a faulty CFTR protein, and so chloride transport is abnormal. Our broad goal in this chapter is to discuss the links between genes and proteins, whether normal or abnormal, and the cause of devastating inherited diseases such as cystic fibrosis.

Three unifying themes will emerge as we follow the processes by which a mutation in a single gene results in a defec-

tive protein and ultimately in a range of defects in the individual organism. The first theme is that information flows from DNA to RNA to protein (Figure 10.2) and from proteins to the building of an organism's phenotype. The gene for the CFTR protein in a healthy person, for example, is normally transcribed into an RNA molecule, which is then translated into CFTR, and this, in turn, regulates normal ion passage through cell membranes, and the person does not suffer the symptoms of cystic fibrosis. A second unifying theme is that protein synthesis requires a large expenditure of energy, and cells have evolved ways that minimize that energy cost by regulating *gene expression*—the translation of genetic information into proteins. A final theme is that basic genetic mechanisms are essentially universal: All creatures, from bacteria to field mice to flowering dogwood, share the same approach to protein synthesis.

This chapter will answer several important questions about how genes function, controlling protein shape and activity and ultimately the activities of living things:

▪ How does information flow from DNA to RNA to proteins?
▪ How can even slight alterations in DNA lead to diseases like cystic fibrosis?
▪ What determines when and where the information in a gene will be used in a cell?

Unifying Chapter Themes

◀ ◀ ◀ ◀ ◀ ◀ ◀ ◀ ◀ ◀ ◀ ◀ ◀ ◀ ◀ ◀ ◀ ◀

Three or four thematic statements in the central example introduce the student to the main topics of the chapter.

Advance Organizer

◀ ◀ ◀ ◀ ◀ ◀ ◀ ◀ ◀ ◀ ◀ ◀ ◀ ◀ ◀ ◀

The chapter's specific objectives are framed as questions to stimulate critical thinking skills.

BOX 9.1   P E R S O N A L   I M P A C T

THE HUMAN GENOME PROJECT

In October of 1990, the journal *Science* included a large foldout wall chart printed with figures that resemble military decorations and candy-striped worms. The poster was, in fact, an up-to-date map of the human genome, placing genes of known function and location, regions where the nucleotide sequence has been worked out, and other genetic data in their correct positions on blown-up, diagrammatic versions of the human's 22 chromosomes plus *X* and *Y* (Figure 1). By looking at the bands of yellow, orange, blue, and magenta, one can interpret how complete—or more accurately, how incomplete—the mapping effort was at that time.

Although molecular geneticists had by then already spent over five years and $100 million of federal research funds in what will eventually be the largest scientific project ever undertaken, there are far more empty spaces on the human gene map than data to plug in. Some have labeled the Human Genome Project the ultimate measure of humankind—an effort that could revolutionize medicine, biology, and psychology. Noting that only a tiny fraction of the map is completed, however, critics wonder whether the benefits can possibly outweigh the tremendous costs.

First suggested in 1985, the Human Genome Project will cost more and take longer than building the atom bomb or landing people on the moon. Hundreds of researchers will labor simultaneously (1) to create a low-resolution physical map that places the 1000 or so currently known human genes in the proper positions on the 23 pairs of chromosomes; (2) to locate and identify the 50,000 to 100,000 additional genes (some 4400 of which occur on an average chromosome); and (3) to determine the nucleotide sequences of all 3 billion bases in the human DNA—even sequences that don't code for proteins or that simply repeat each other.

The project will cost at least $3 billion and won't be completed before the year 2005. The result will be a 500-volume encyclopedia of human genetics spelling out the detailed instructions for every protein in the body, and with them, a high-resolution view of how these proteins function normally day to day and perhaps participate in mental and physical illness. Collectively, humans have more than 4000 hereditary diseases, including sickle-cell anemia, cystic fibrosis, and Huntington's disease. These three and a few hundred others can be traced to single-gene mutations. Most diseases, however, including cancer, diabetes, Alzheimer's, and heart disease, probably have roots in multiple genes, and the interactions between them will be challenging to work out, even with a genome map.

To some critics—including 23 Harvard professors who published a dissenting view of the Human Genome Project in *Science* shortly before the 1990 map appeared—the promised high-resolution view is a mixed blessing. Knowing the complete nucleotide sequence, wrote Bernard Davis and colleagues, would be "like viewing a painting through a microscope," with researchers having to "plow through 1 to 2 million 'junk' bases

FIGURE 1 ■ The Human Genome Map.

before encountering an interesting sequence"* and then trying to discover its unknown function. Even senior statesman of molecular biology Sydney Brenner (see Box 7.2 on page 155) jokes that working out nucleotide sequences is so tedious that a penal colony could hand out sequencing projects as prison sentences. And everyone from the critics to Nobelist James Watson, who is heading those parts of the project overseen by the National Institutes of Health, agree that only 3 percent of the bases in the total sequence code for proteins of any kind.

Other major criticisms are financial and ethical. The last decade or so has seen major cuts in funding for biomedical research. The money allocated for Genome Project work in 1991 alone could provide sizable ($200,000) research grants to over 750 investigators. Critics also worry that a human gene map could lead to genetic discrimination, since doctors, insurance companies, and employers could "read" your inherited tendencies toward heart disease, cancer, or dementia, let's say, even though you have no symptoms now and may never have them.

Despite these legitimate concerns, many scientists are convinced that the benefits will far exceed the risks and point to the wealth of basic information already unlocked from DNA sequencing in viruses, bacteria, plants, and other animals. James Watson, probably the Genome Project's most eloquent spokesman, sees it this way: "When finally interpreted, the genetic message encoded within our DNA molecules will provide the ultimate answers to the chemical underpinnings of human existence."†

*Bernard D. Davis et al., "The Human Genome and Other Initiatives," *Science* 249 (1990): 342–343.

†James D. Watson, "The Human Genome Project: Past, Present, and Future," *Science* 248 (1990): 44–49.

Boxed Essays

◄   ◄   ◄   ◄   ◄   ◄   ◄   ◄   ◄   ◄   ◄

These stimulating accounts explore the scientific search for answers, the people behind the results, and how biology influences our own lives and our environment. There are four kinds of boxes: How Do We Know?; Biology: A Human Endeavor; Personal Impact; and Focus on the Environment.

Emphasis on the Environment

▶   ▶   ▶   ▶   ▶   ▶   ▶   ▶   ▶   ▶   ▶   ▶   ▶   ▶   ▶

An increased coverage of environmental science helps the student make a connection between the principles of biology and the world in which he or she lives. Here, both text and photo demonstrate how anaerobic respiration relates to the balance of nature.

FIGURE 5.11 ■ **Environmental Carbon Cycling: Anaerobic Respiration Plays a Key Role.** Hikers pausing for a rest by a clear mountain lake see only the sparkling upper layers of water, where photosynthesis in the tissues of algae and green plants fixes carbon and releases oxygen into the water. When water plants and oxygen-breathing fish die, their bodies sink to the lake bottom. These deep layers are colder and soon become depleted of oxygen, as oxygen-utilizing microorganisms begin to decompose the debris. Once the oxygen is fully depleted, the microorganisms switch to anaerobic metabolism and complete the decomposition of the organic matter to carbon dioxide and often methane gas, which hikers can sometimes see bubbling up to the surface. Without this anaerobic recycling, most of the carbon would be tied up in dead plants and animals.

During fermentation in muscle cells, yeast cells, and certain other kinds of microorganisms, the pyruvate molecules are degraded to wastes such as alcohol, carbon dioxide, and lactic acid, while the NADH is recycled to NAD$^+$. The energy yields of these two metabolic pathways may seem meager—just two ATPs per glucose. However, anaerobic metabolism (energy breakdown in the absence of oxygen) is crucial to the global recycling of carbon and the stability of the environment.

Organic matter from dead leaves, dead microorganisms, and other sources often sinks into an environment devoid of oxygen, such as the soft layers at the bottom of lakes or oceans (Figure 5.11). If it weren't for anaerobic decomposers—organisms capable of breaking down organic matter via anaerobic metabolism—most of the world's carbon would eventually be locked up in undecomposed organic material in these oxygen-poor environments, and deeper and deeper layers would build up. As a result, there would be too little carbon dioxide available as a raw material for photosynthesis, plants would be unable to generate new glucose molecules, and neither plants nor animals would survive.

Although simple and microscopic, anaerobic decompos-

ers clearly play an enormous role in the balance of nature. The vast majority of life forms, however, expire fairly quickly without oxygen. But their dependence on oxygen is based on a form of metabolism that provides an energy harvest 18 times greater than that provided by anaerobic metabolism.

AEROBIC RESPIRATION: THE BIG ENERGY HARVEST

While muscle cells as well as yeasts and other decomposers have the ability to metabolize sugars even when oxygen is unavailable (for a while, at least), the vast majority of living things, from euglenas to elephants, roses to redwoods, are made up of cells that require oxygen for their major metabolic pathway. Known as aerobic respiration, this pathway shunts the products of glycolysis through the Krebs cycle and harvests energy in the electron transport chain. Being 18 times more productive than glycolysis alone, this extended pathway can provide the huge amounts of ATP an active cell needs, and its superiority is clearly demonstrated: One out of every 500 yeast cells has a mutation, or change, in its DNA that prevents it from carrying out aerobic respiration even when oxygen is present. These so-called *petite* yeasts get by solely on glycolysis and fermentation, but they grow two to three times more slowly than their normal counterparts.

The reason for the greater efficiency of aerobic respiration is revealed by the overall equation for aerobic respiration:

$$C_6H_{12}O_6 + 6 O_2 + 36 ADP + 36 P_i \rightarrow 6 CO_2 + 6 H_2O + 36 ATP$$

The initial glucose molecule is broken down completely to inorganic waste products that are not energy-rich; thus, most of the energy residing in the molecular bonds of the sugar is released and a large proportion of it stored as ATP—36 ATPs per glucose, to be precise.

Aerobic respiration has two phases that use the products of glycolysis. The first phase is known as the Krebs cycle, named after the scientist who worked out its reactions, Sir Hans Krebs. The **Krebs cycle** of reactions cleaves the carbons from pyruvate, releases them as carbon dioxide, and stores energy in reduced carrier molecules. The second phase, the **electron transport chain,** then strips the electrons from the reduced carriers (Figure 5.12). The flow of electrons down this transport chain creates a current that is then used to build the cellular currency ATP.

The Krebs Cycle: Metabolic Clearinghouse

As in fermentation, the two pyruvate molecules produced from each glucose during glycolysis are the raw material for the Krebs cycle. Unlike fermentation, however, the important events of the Krebs cycle take place not in the cytoplasm but in the mitochondria of eukaryotic cells (or on the plasma membranes of bacteria). Recall from Chapter 3 that the mito-

Complex Topics Made Relevant

▶ ▶ ▶ ▶ ▶ ▶ ▶ ▶ ▶ ▶ ▶ ▶ ▶ ▶ ▶ ▶ ▶ ▶ ▶

Many topics, such as cell and molecular biology, are made relevant through the use of health and ecological applications. Here, both text and photo demonstrate the relationship between the endoplasmic reticulum and human health.

▶ ▶ ▶

FIGURE 3.18 ■ **Some Bedouin Women Have a Smooth ER Problem.** Because this woman's clothing leaves little or no skin exposed to sunlight, her smooth ER may not be able to make enough of the vitamin D necessary to maintain strong, healthy bones.

African women of one Bedouin tribe who wear dark, full-length garments get very little exposure to sunlight. As a result, the regions of smooth ER in their cells are often unable to convert enough cholesterol to vitamin D, and their bones grow soft and weak (Figure 3.18).

Another part of the endoplasmic reticulum is studded with ribosomes, looks rough under the microscope, and is called the **rough ER.** This region is involved in the synthesis of certain proteins. For example, the rough ER helps produce the enzymes that digest ice cream and most of the other foods you eat. Tracing the production and transport of an enzyme through this membrane system helps us understand how the ER functions. Let's focus on the rough ER in cells within a person's pancreas, a cucumber-shaped organ that generates digestive enzymes and secretes them into a duct leading to the intestine, where food is broken down and digested.

Within a pancreatic cell, the nucleus makes a special RNA called *messenger RNA* that carries genetic information (in this case, information for the digestive enzyme) out of the nucleus and into the cytoplasm (see Figure 3.15). Once in the cytoplasm, the messenger RNA joins the small, beadlike ribosomes—biochemical anvils on which protein molecules will be forged. The ribosomes then stick to the surface of the rough ER—the reason, in fact, that it appears rough. Proteins like these digestive enzymes that are assembled on rough ER enter the ER cavity, are modified as they move along through the channels, and are eventually pinched off in little sacs, or vesicles.

Most of the sacs pinched off from the endoplasmic reticulum enter another membrane system, the **Golgi apparatus,** where the digestive enzymes or other proteins in the sacs are

further modified. The modified proteins then leave the Golgi apparatus headed either for the cell's plasma membrane, for tiny digestive sacs within the cell (the lysosomes), or for export from the cell (see Figure 3.17). When exported from the pancreatic cells, the digestive enzymes move down a duct and into the intestine. There they can break down the fats, sugars, and other nutrients present in the ice cream.

The Golgi apparatus is a key component of the membranous organelles along the enzyme secretion pathway. Named after the Italian cell biologist who first spotted it, the Golgi apparatus is like a packaging department for the eukaryotic cell. In 1898, Camillo Golgi observed this apparatus in the nerve cells of a barn owl, but it was not until the availability of electron microscopes over 40 years later that biologists could see the structure clearly. Each Golgi apparatus is a stack of saucer-shaped, baglike membranes surrounded by small, round, membranous containers, or **vesicles** (see Figure 3.17c). A cell can have just one Golgi apparatus or many thousands. Golgi-packaged proteins and lipids repair the plasma membrane itself when it is damaged; and in plants, the Golgi apparatus (called a *dictyosome* by botanists) packages for export the precursors to the cellulose that forms the outer cell wall.

■ **Vesicles in the Cytoplasm** While many vesicles that pinch off from a Golgi apparatus leave the cell, two main types of vesicles—lysosomes and microbodies—take up permanent residence in the cytoplasm. **Lysosomes** are spherical vesicles

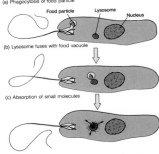

FIGURE 3.19 ■ *Euglena* **Engulfing a Food Particle.** (a) Phagocytosis. When a *Euglena* cell happens upon a food particle of the right size and composition, the cell membrane forms a pocket around it and engulfs it. (b) Fusion. A lysosome then fuses with the food vacuole, and digestive enzymes break down the food particle. (c) Absorption. Small nutrient molecules then pass through the lysosome membrane into the cytoplasm and nourish the cell.

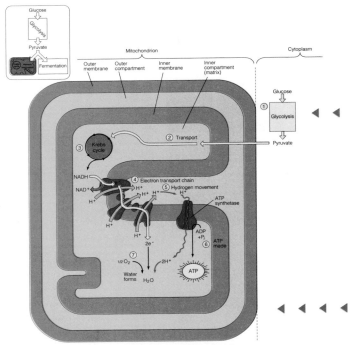

FIGURE 5.16 ■ **Mitochondrion: Site of the Krebs Cycle and Electron Transport Chain.** This diagram shows where the phases of aerobic respiration take place in the mitochondrion. Pyruvate molecules (1) generated during glycolysis in the cytoplasm are transported through the outer and inner mitochondrial membranes to the inner compartment (matrix) (2), where the Krebs cycle takes place (3). Then electrons from energy carriers are passed down the electron transport chain (4), and proteins use this released energy to pump hydrogen ions across the inner membrane (5). Finally, these hydrogen ions flow through the protein ATP synthetase back across the inner mitochondrial membrane, and the protein traps energy in ATP (6) much as a turbine captures the energy of water flowing over a dam and converts it to electricity. The final acceptor of the electrons is oxygen (7), which joins hydrogen ions and electrons and forms water.

Extensive Use of Icons

◀ ◀ ◀ ◀ ◀ ◀ ◀ ◀ ◀ ◀ ◀ ◀ ◀ ◀ ◀ ◀ ◀ ◀ ◀

Orienting diagrams place structures and processes into a physical context for the student.

Extensive Use of Process Diagrams

◀ ◀ ◀ ◀ ◀ ◀ ◀ ◀ ◀ ◀ ◀ ◀

These figures depict sequential biological events with individual steps numbered and keyed to step-by-step discussions in the text or figure legend.